TORQUE CONCEPT VISUALS
George Mathew

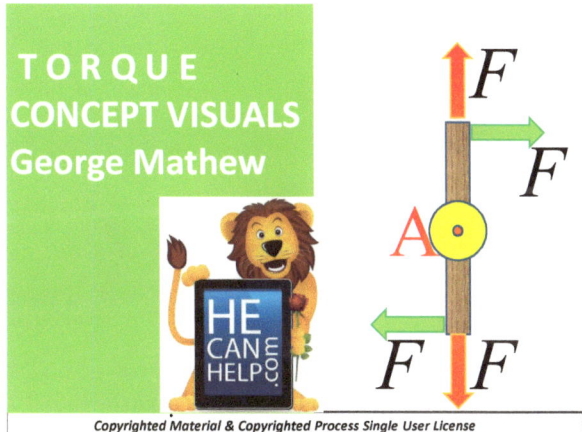

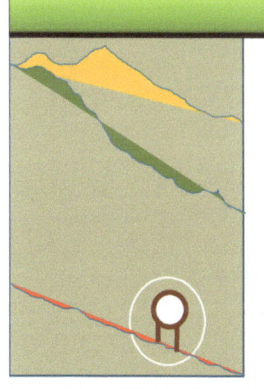

Copyrighted Material & Copyrighted Process Single User License

COLLEGEPHYSICS
Direction of
Net TORQUE

©

George Mathew, Ph. D.

info@**hecanhelp**.com

Published by

ACADEMICS ON LINE INTL

PO Box 1324, *Mason*

OH 45040

USA

George Mathew, Ph. D.

INFO@HECANHELP.com

Single User License

© *Copyrighted*
Material
& Copyrighted
Process

George Mathew, Ph. D.

INFO@hecanhelp.com

Problems 1-25:

What is the net torque

on the system with respect to

axis of rotation at the point A?

A. Net torque is clockwise.

B. Net torque is counter – clockwise.

C. Net torque is zero.

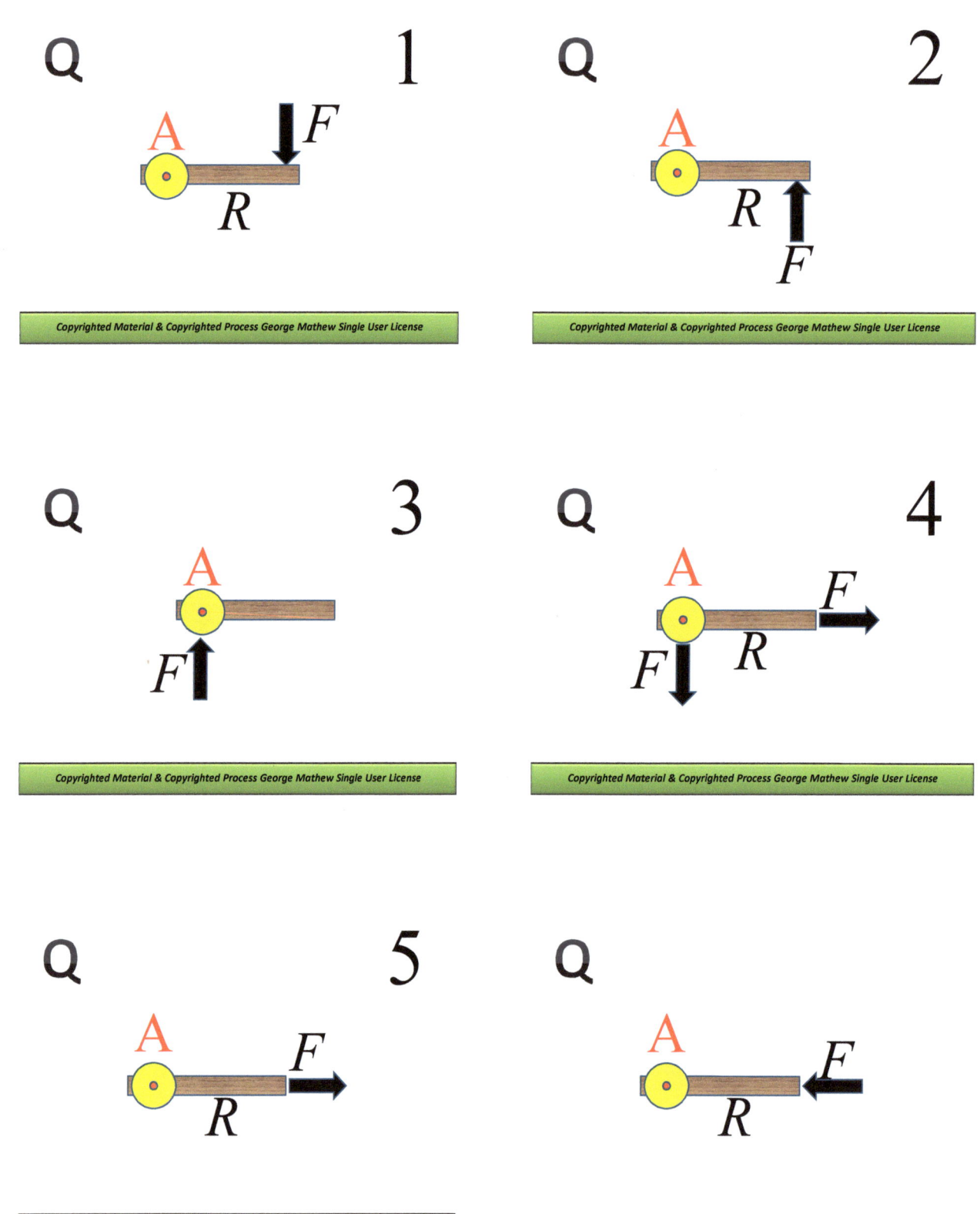

Q 1

Q 2

Q 3

Q 4

Q 5

Q

Q 7

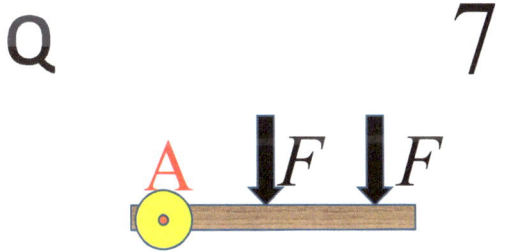

Q 8

A

F

R

F

Q 9

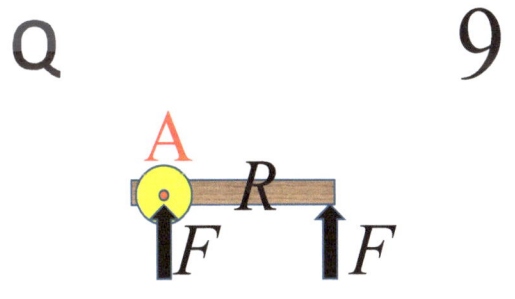

Q 10

A

F

R F

Problems 1-25:

What is the net torque on the system with respect to axis of rotation at point A?

A. *Net torque is clockwise.*

B. *Net torque is counter – clockwise.*

C. *Net torque is zero.*

Q 11

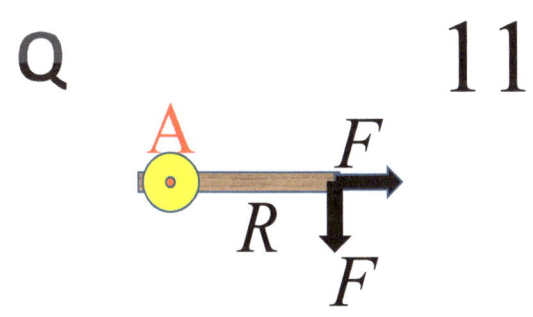

Q 12

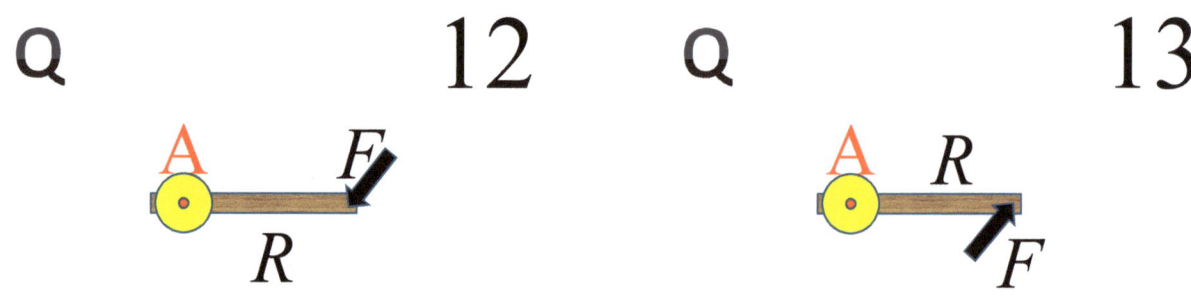

Q 13

Q 14

Q 15

Q 16

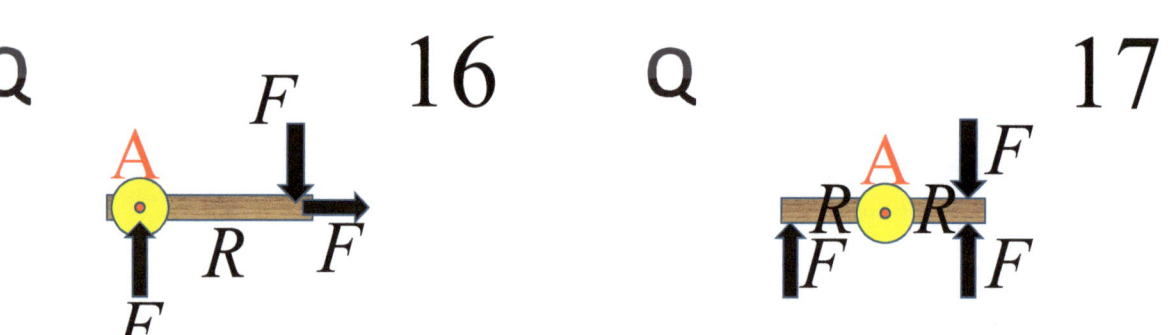

Q 17

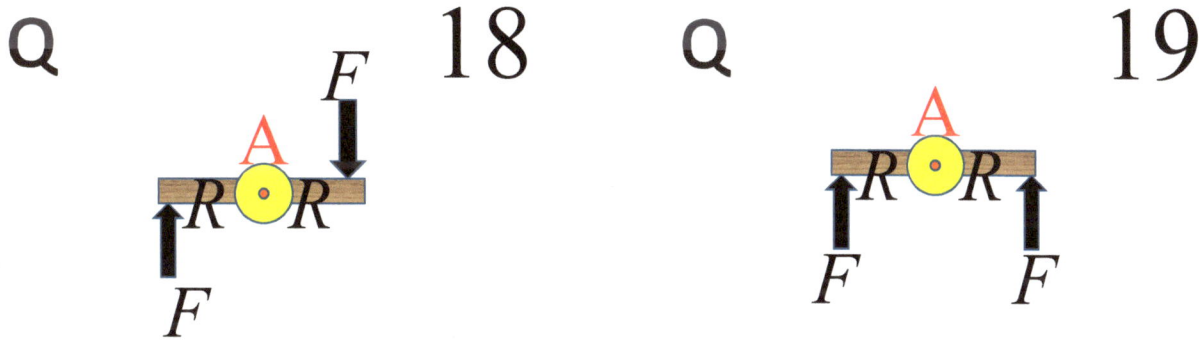

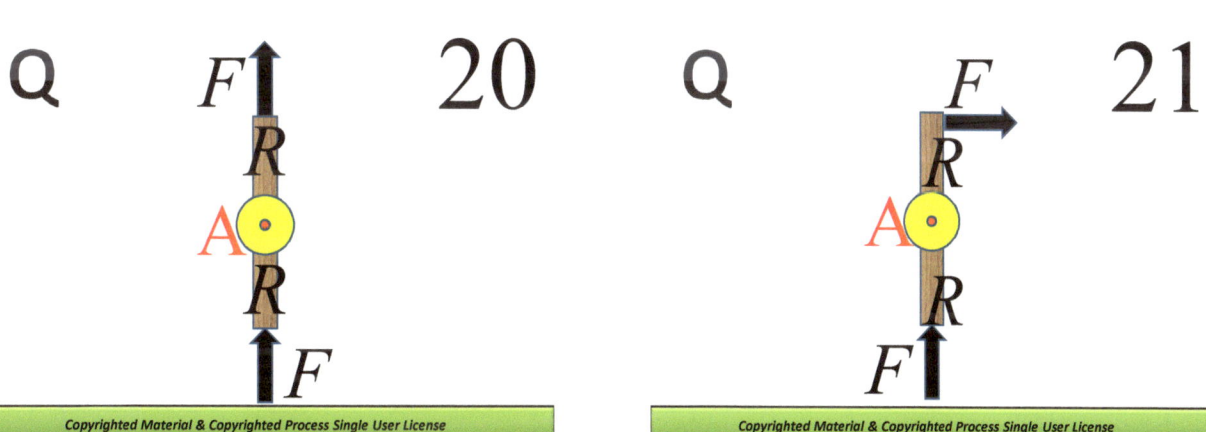

Q 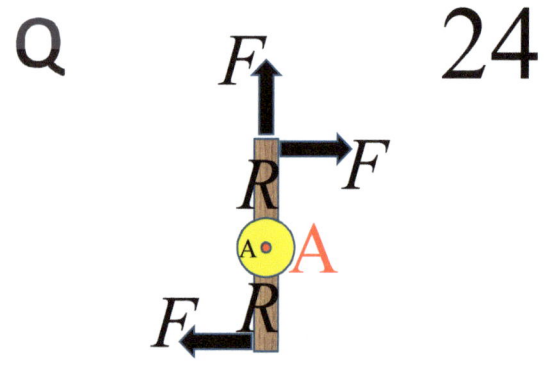 24

Q 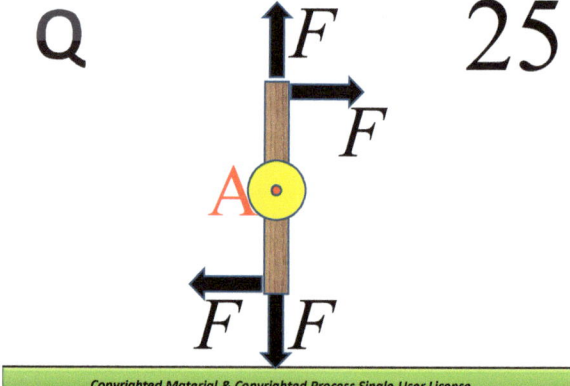 25

FORCES IN green
PRODUCE CLOCKWISE TORQUES
FORCES IN red
PRODUCE zero TORQUE
FORCES IN blue
PRODUCE counter-CLOCKWISE TORQUES

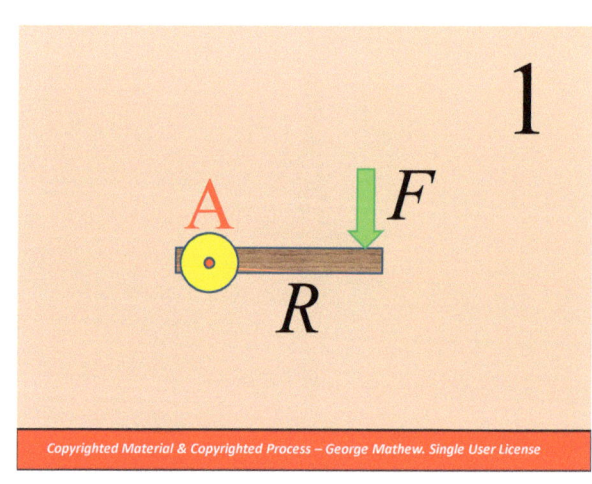

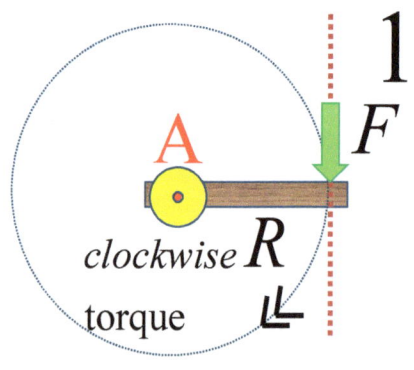

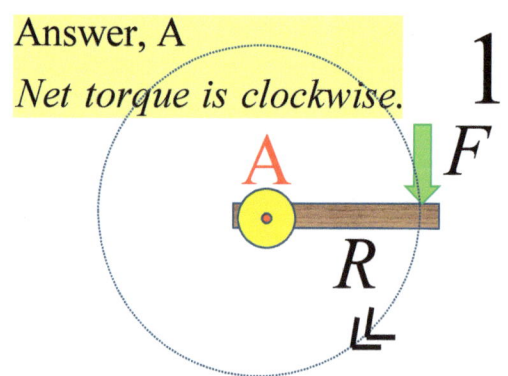

Answer, A
Net torque is clockwise.

FORCES IN green
PRODUCE CLOCKWISE TORQUES
FORCES IN red
PRODUCE zero TORQUE
FORCES IN blue
PRODUCE counter-CLOCKWISE TORQUES

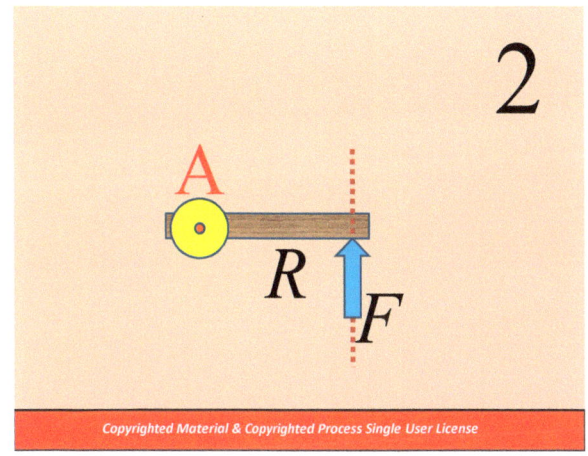

Net Torque
is conter-clockwise.

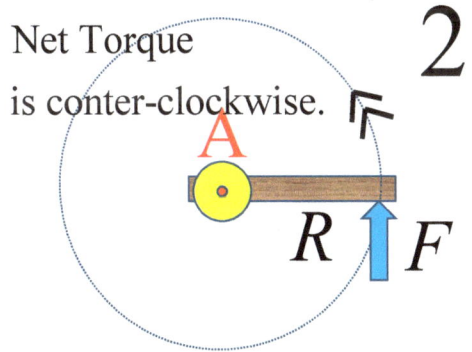

FORCES IN green
PRODUCE CLOCKWISE TORQUES
FORCES IN red
PRODUCE zero TORQUE
FORCES IN blue
PRODUCE counter-CLOCKWISE TORQUES

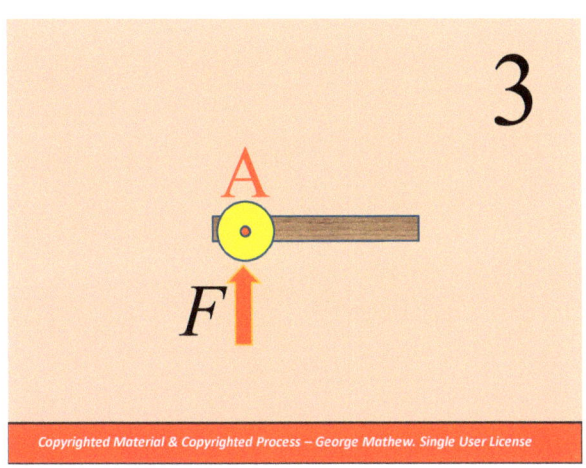

Torque is zero.
Line of action of the force
pass through the axis of rotation.
(Zero perpendicular distance.)

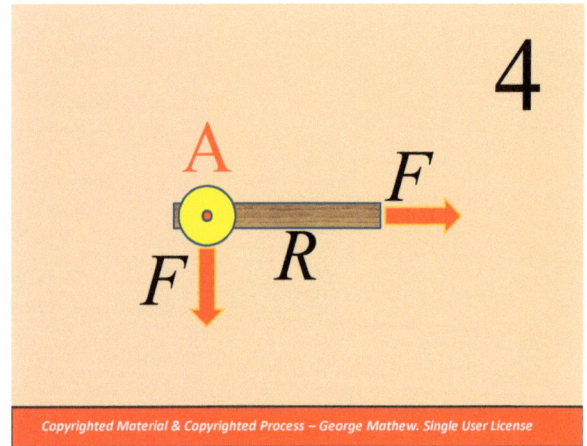

We see two forces.

Take torque of EACH force.

Separate into clockwise and counter-clockwise torques.

Add clockwise.

Add counter-clockwise.

Find net torque.

1. Torque is zero.

Line of action of the force pass through the axis of rotation. (Zero perpendicular distance.)

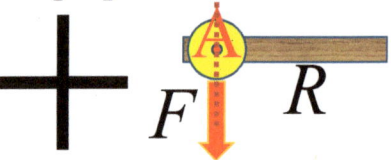

2. Torque is zero.

Line of action of the force pass through the axis of rotation. (Zero perpendicular distance.)

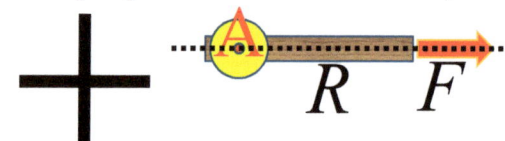

Net torque is zero.

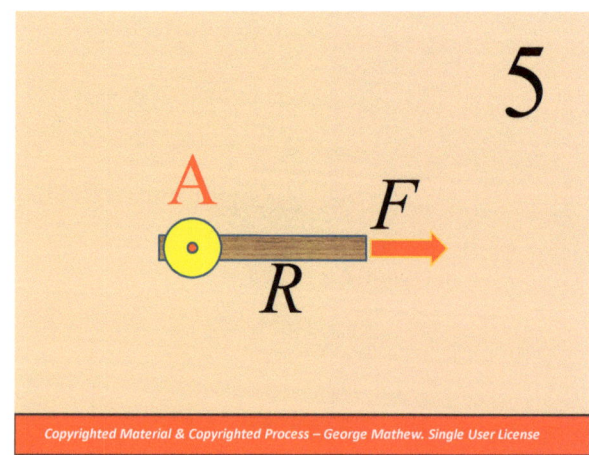

Torque is zero.

Line of action of the force

pass through the axis of rotation.

(Zero perpendicular distance.)

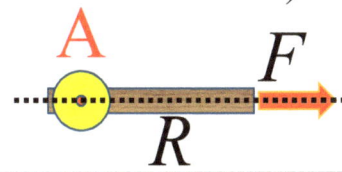

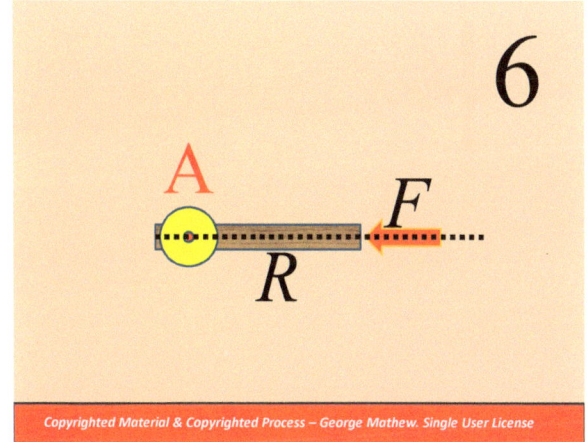

Torque is zero.

Line of action of the force

pass through the axis of rotation.

(Zero perpendicular distance.)

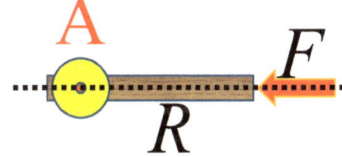

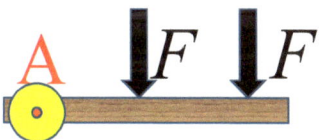

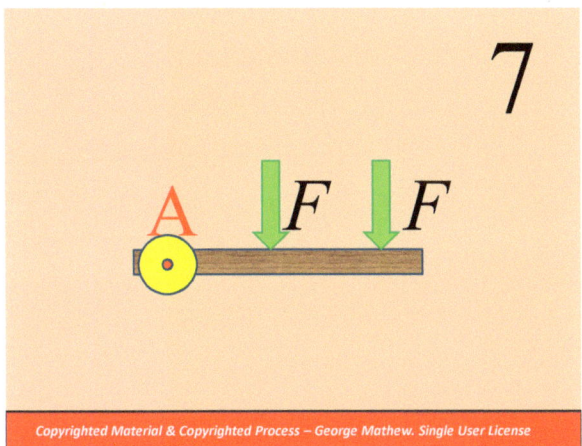

We see two forces.

Take torque of EACH force.

Separate into clockwise and

counter-clockwise torques.

Add clockwise.

Add counter-clockwise.

Find net torque.

7

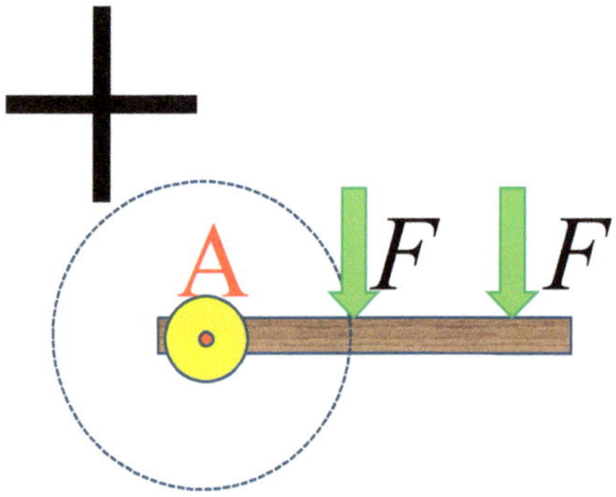

7

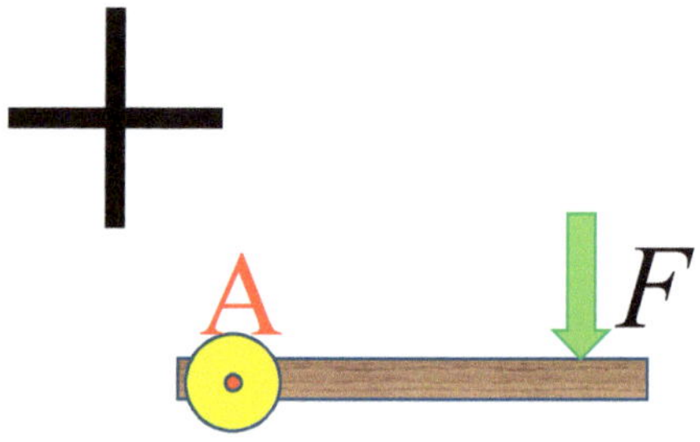

Answer, A

Net torque is clockwise.

7

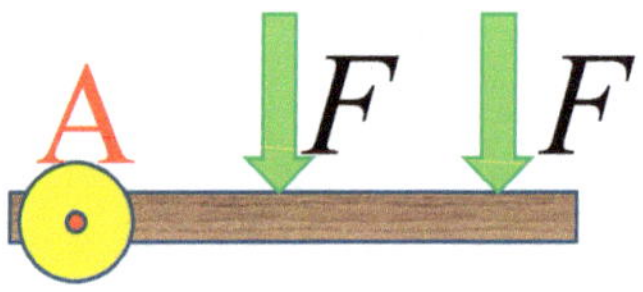

8

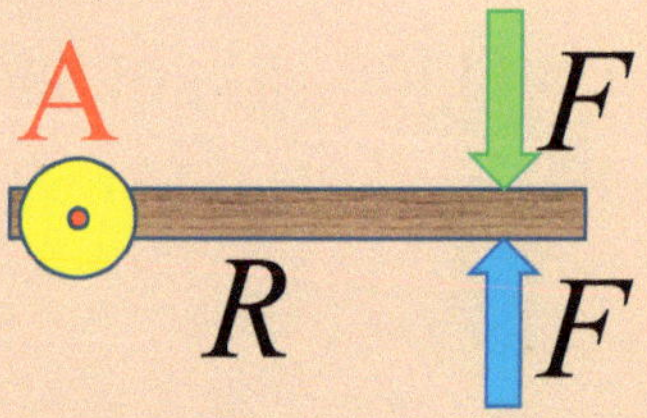

8

We see two forces.

Take torque of EACH force.

Separate into clockwise and counter-clockwise torques.

Add clockwise.

Add counter-clockwise.

Find net torque.

8

Two equal forces acting in opposite directions at the same point will cancel each other's torque! Net torque = 0.

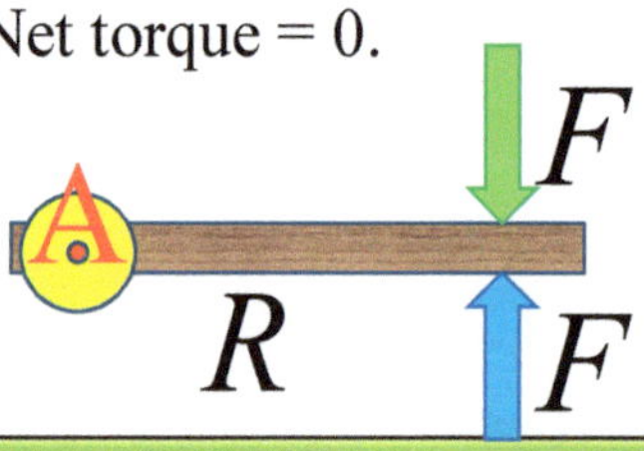

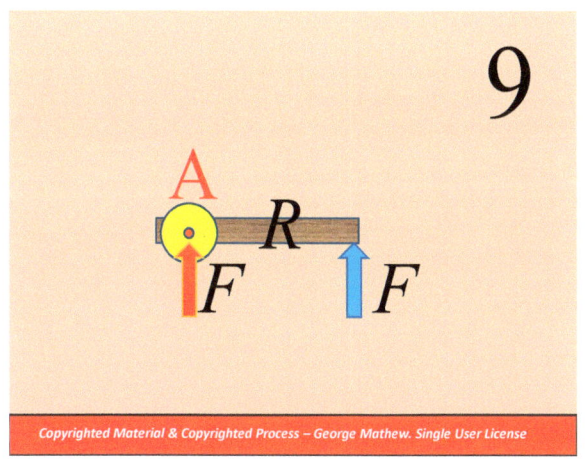

9

We see two forces.
Take torque of EACH force.
Separate into clockwise and
counter-clockwise torques.
Add clockwise.
Add counter-clockwise.
Find net torque.

9

9

9

counter –
clockwise
torque F

Net Torque is
counter-clockwise.

9

10

We see two forces.

Take torque of EACH force. 10

Separate into clockwise and

counter-clockwise torques.

Add clockwise.

Add counter-clockwise.

Find net torque.

10

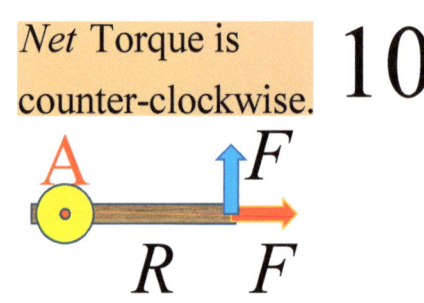

F

10

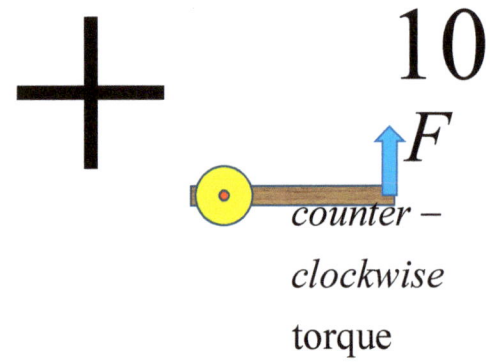

F

counter –

clockwise

torque

Net Torque is

counter-clockwise. 10

A *F*

R *F*

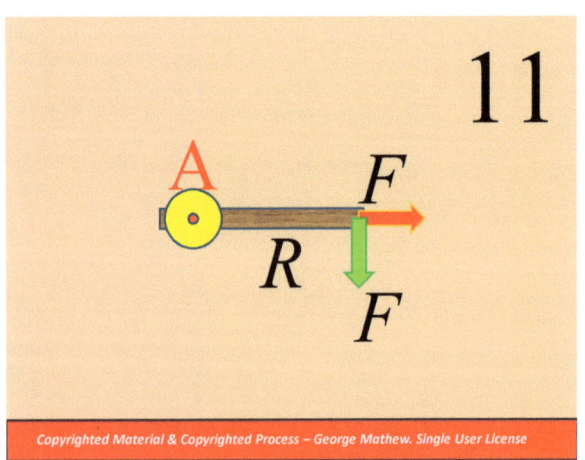

11

We see two forces.

Take torque of EACH force. 11

Separate into clockwise and

counter-clockwise torques.

Add clockwise.

Add counter-clockwise.

Find net torque.

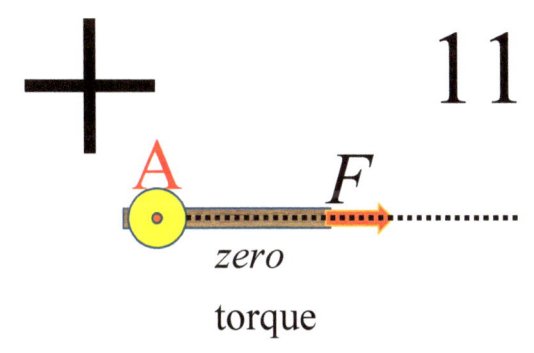

11

zero

torque

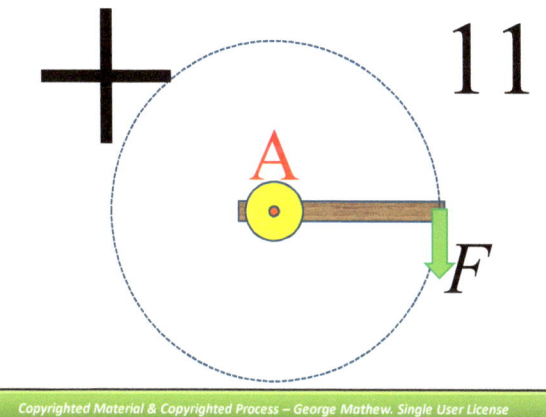

11

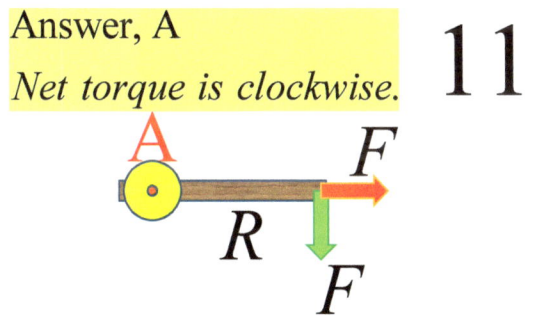

Answer, A

Net torque is clockwise.

11

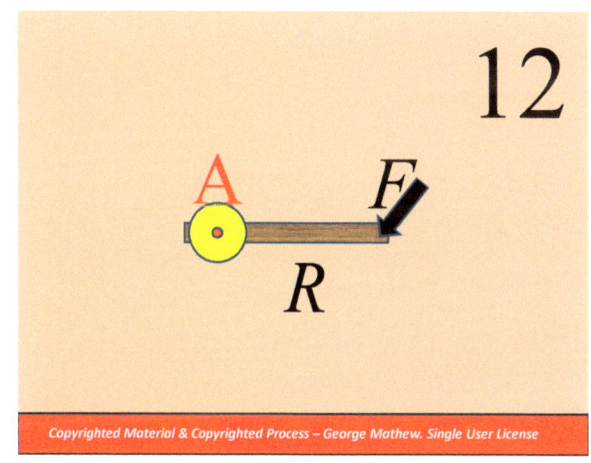

12

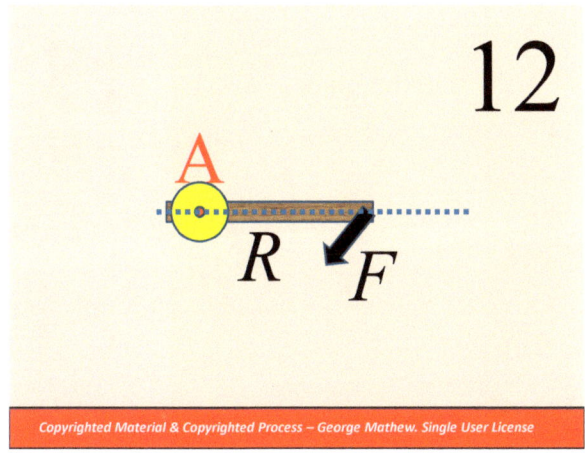

12

12

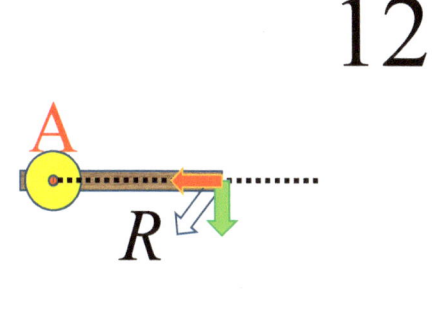

12

We see two force components. 12
Take torque of EACH force.
Separate into clockwise and
counter-clockwise torques.
Add clockwise.
Add counter-clockwise.
Find net torque.

12

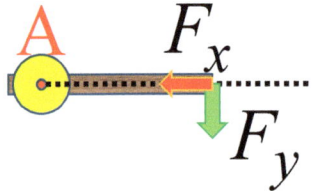

torque

12

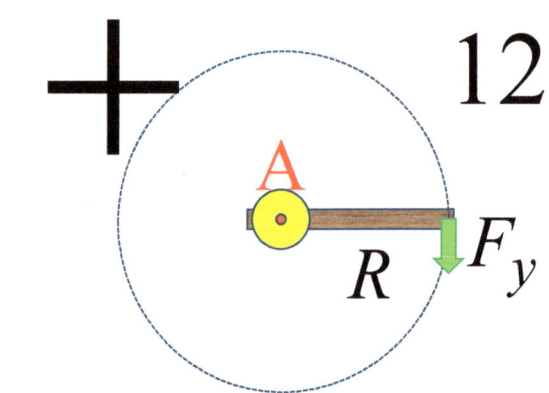

Answer, A 12
Net torque is clockwise.

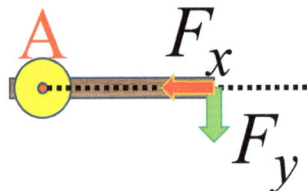

13

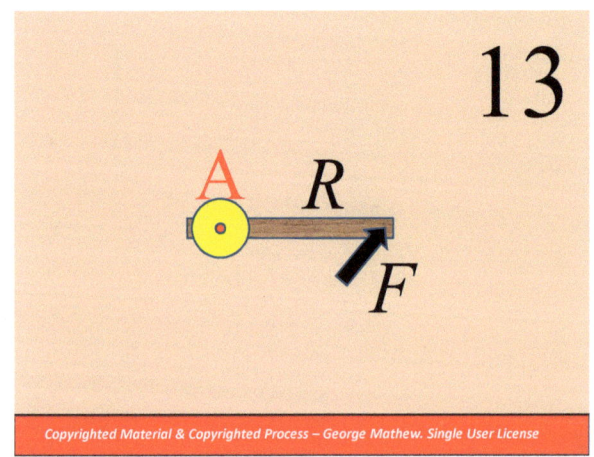

13

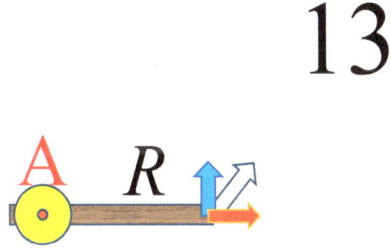

13

A F_y

F_x

We see two force components.
Take torque of EACH force.
Separate into clockwise and
counter-clockwise torques.
Add clockwise.
Add counter-clockwise.
Find net torque.

13

13

zero
torque
F_x

13

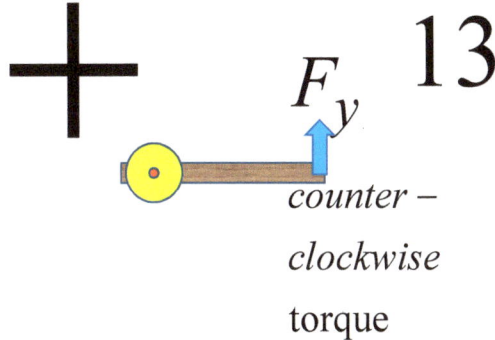

*counter –
clockwise*
torque

13

Net torque is
counter-
clockwise.

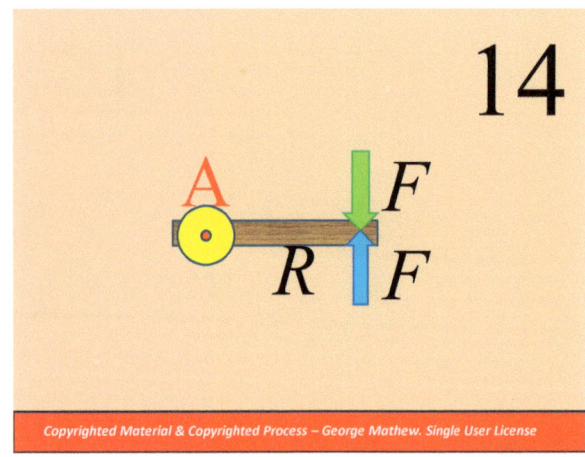

14

Net Torque is zero.

14

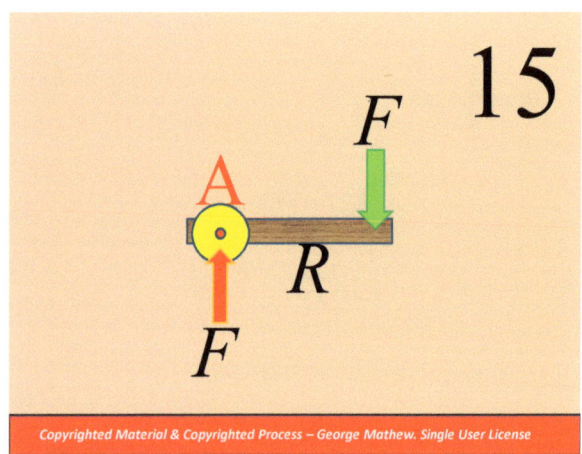

15

We see two forces.
Take torque of EACH force.

15

Separate into clockwise and
counter-clockwise torques.
Add clockwise.
Add counter-clockwise.
Find net torque.

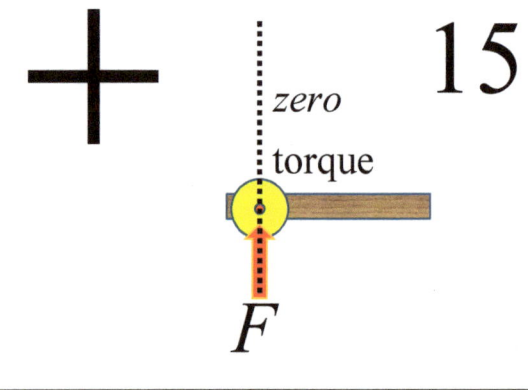

15

zero
torque

F

15

clockwise
torque

Net torque is clockwise.

15

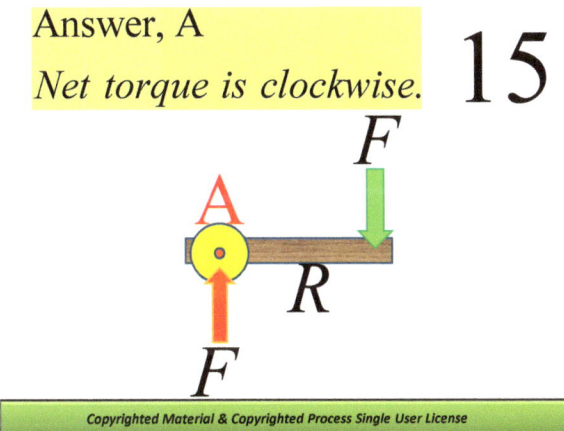

16

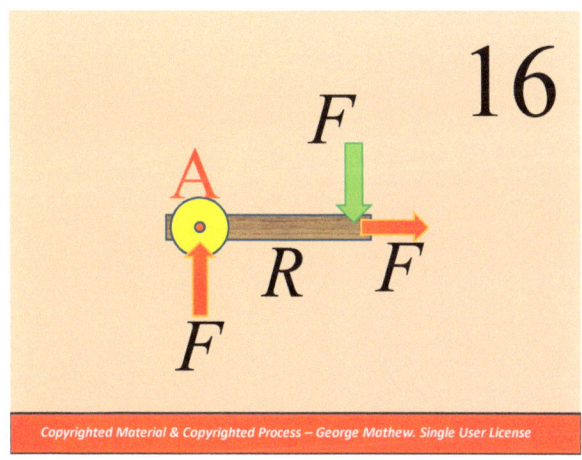

We see three forces.

Take torque of EACH force.

Separate into clockwise and

counter-clockwise torques.

Add clockwise.

Add counter-clockwise.

Find net torque.

16

16

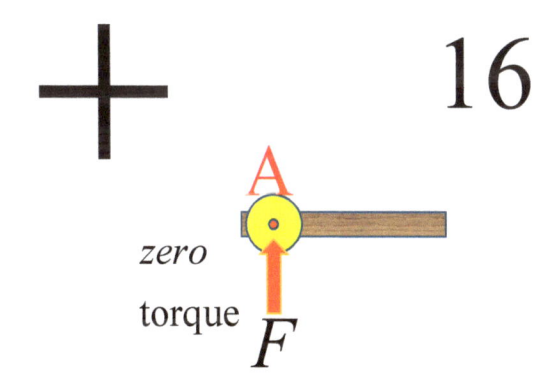

zero

torque

16

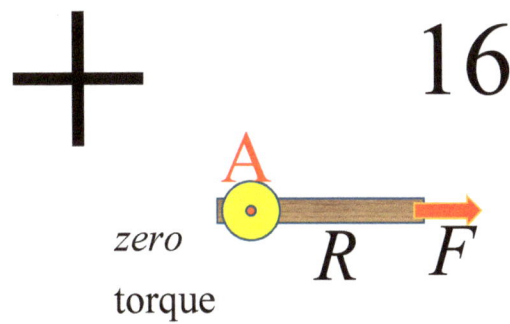

zero

torque

16

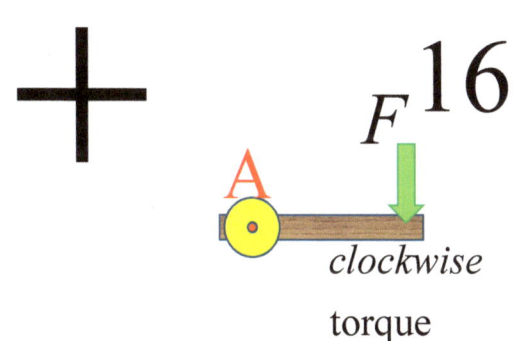

clockwise

torque

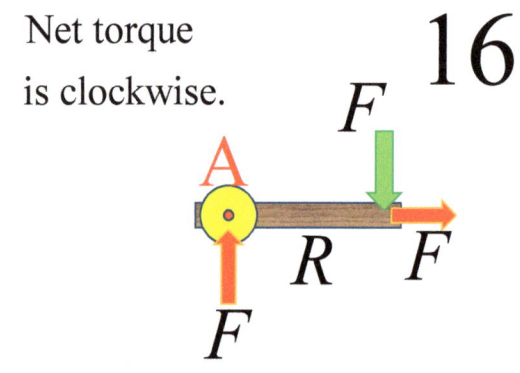

Net torque is clockwise.

16

17

We see three forces.

Take torque of EACH force.

Separate into clockwise and counter-clockwise torques.

Add clockwise.

Add counter-clockwise.

Find net torque.

Torques of these two forces about the same axis of rotation cancel to zero.

17

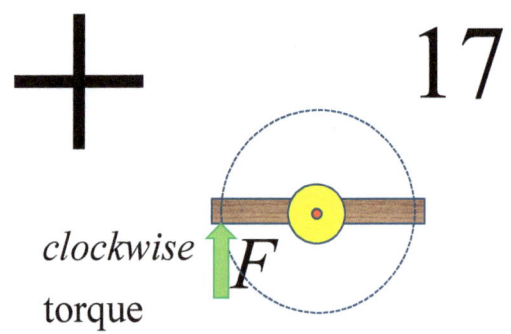

17

clockwise torque

Answer, A
Net torque is clockwise.

17

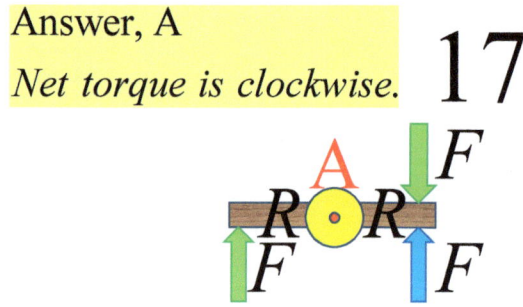

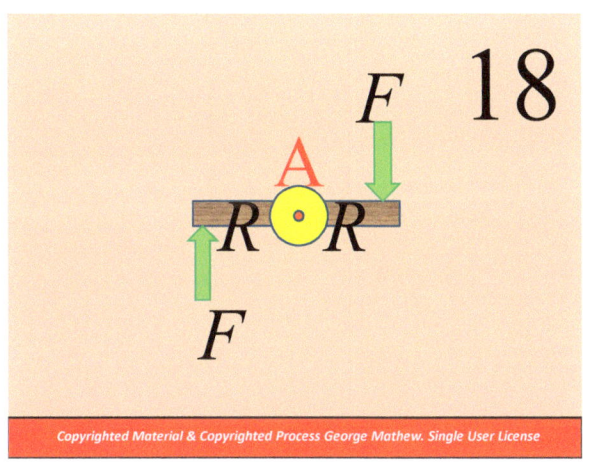

18

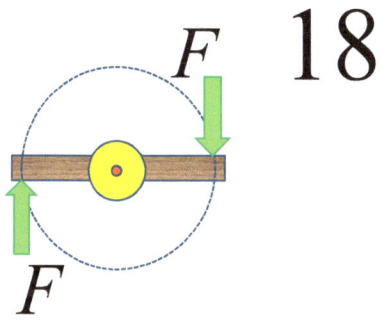

18

We see two forces.

Take torque of EACH force.

Separate into clockwise and

counter-clockwise torques.

Add clockwise.

Add counter-clockwise.

Find net torque.

18

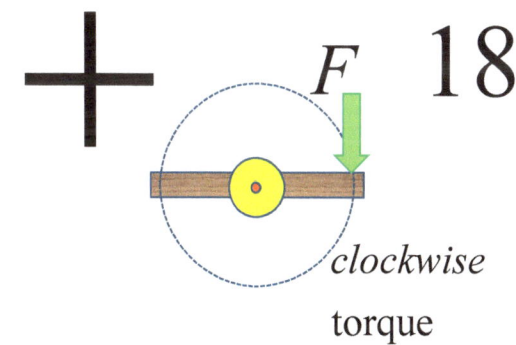

clockwise

torque

18

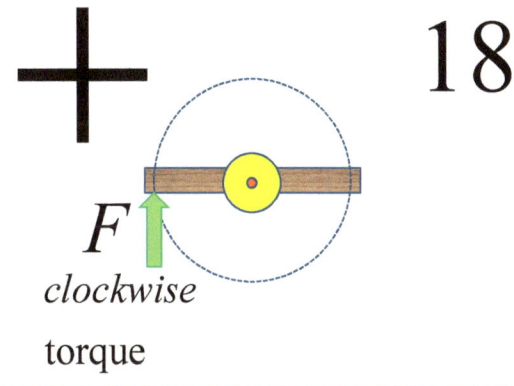

clockwise

torque

18

Answer, A

Net torque is clockwise.

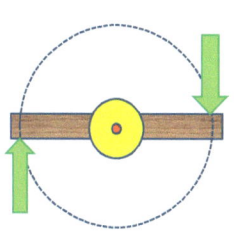

18

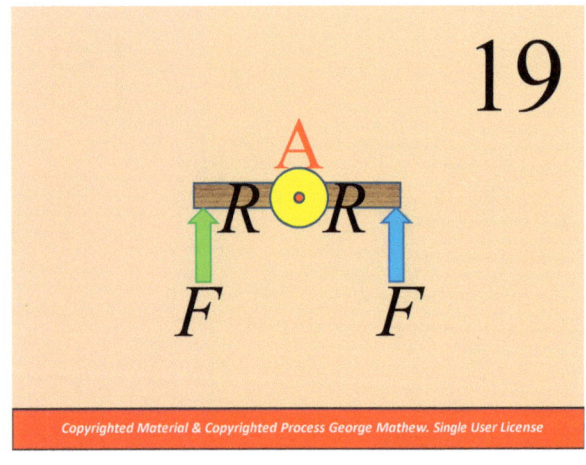

19

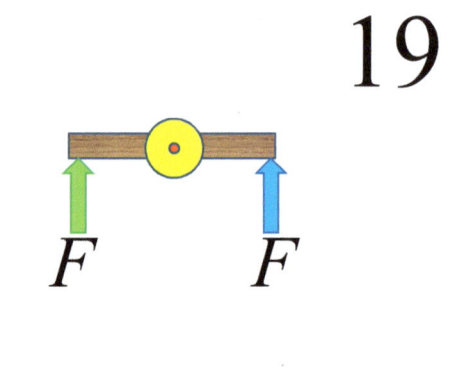

19

We see two forces.

Take torque of EACH force. 19

Separate into clockwise and

counter-clockwise torques.

Add clockwise.

Add counter-clockwise.

Find net torque.

+ 19

counter − clockwise

torque

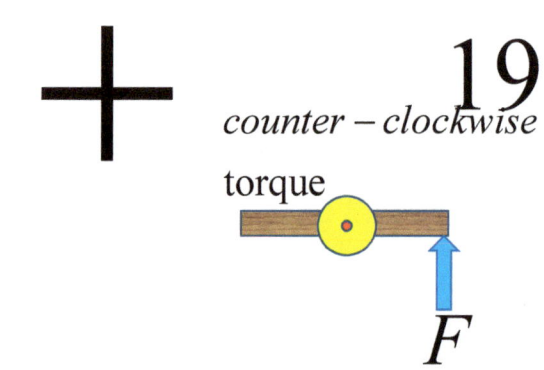

F

+ 19

clockwise

torque F

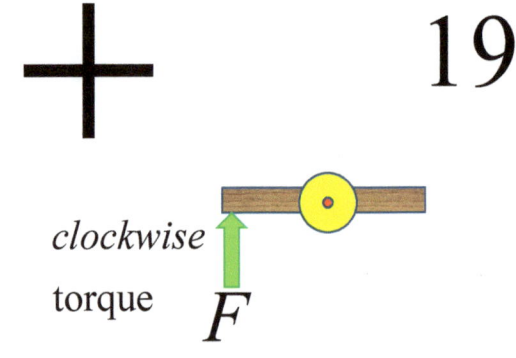

19

Net torque is zero.

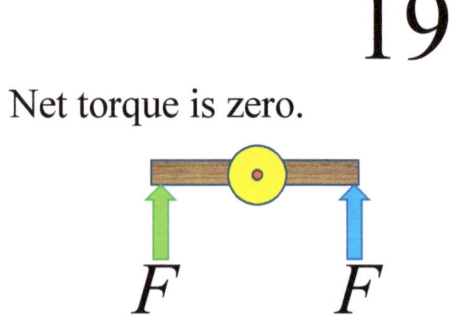

F F

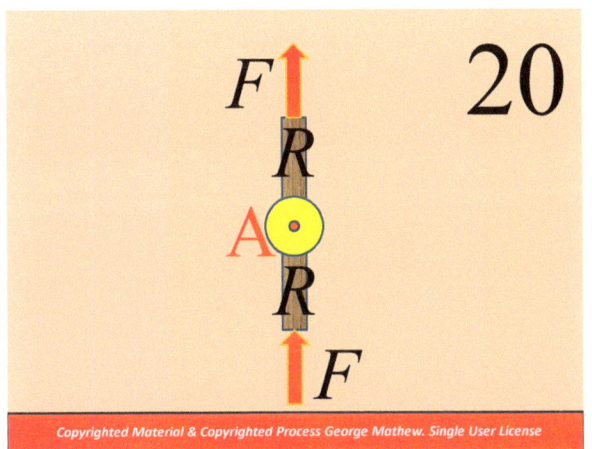

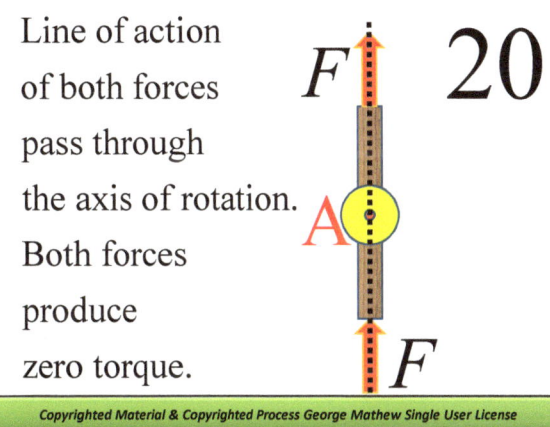

Line of action of both forces pass through the axis of rotation. Both forces produce zero torque.

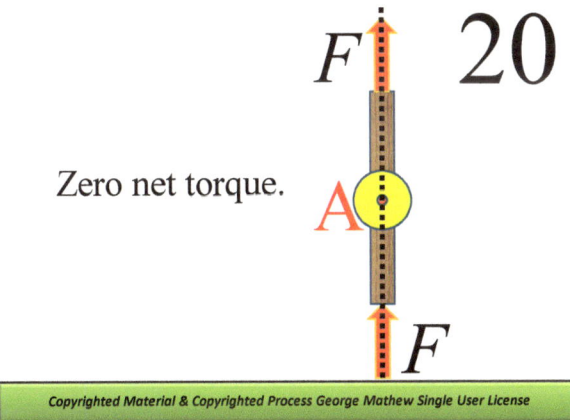

Zero net torque.

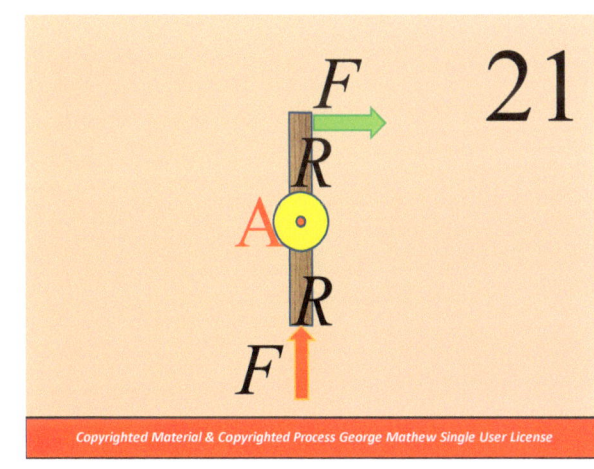

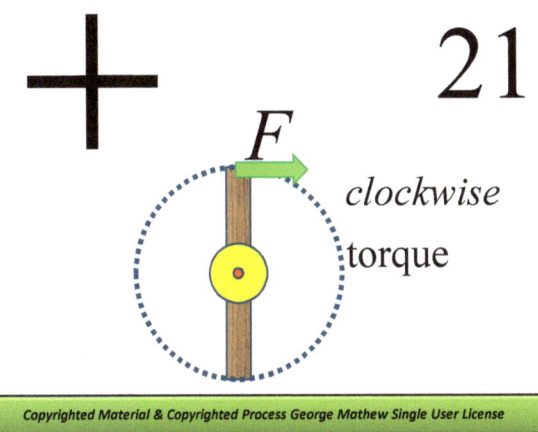

clockwise torque

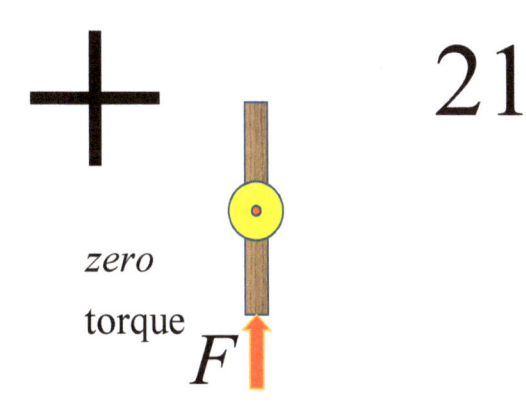

zero torque

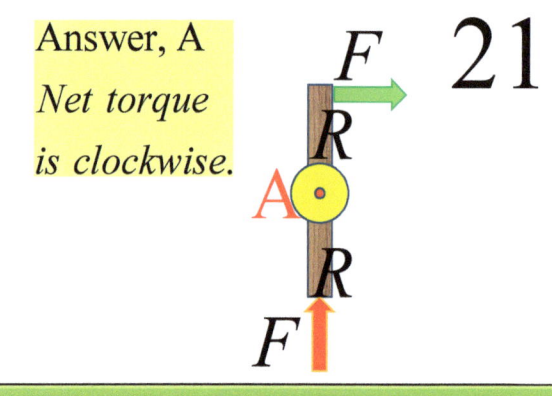

21

Answer, A
Net torque is clockwise.

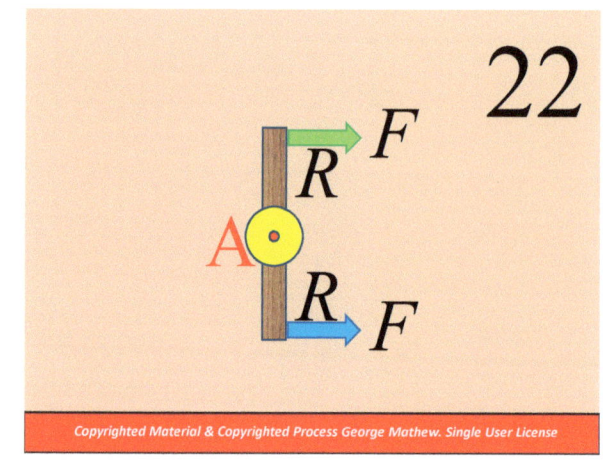

22

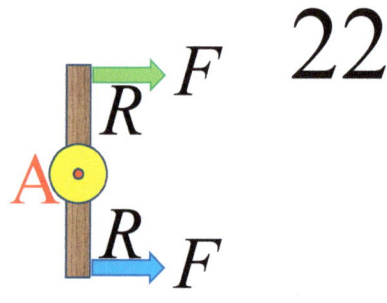

22

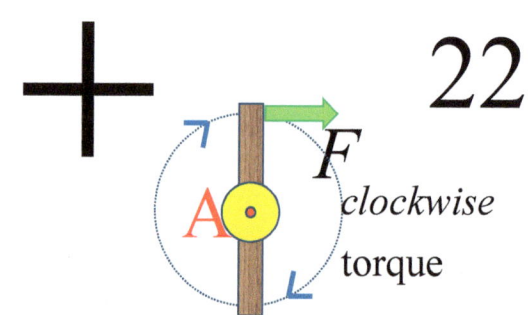

22

clockwise torque

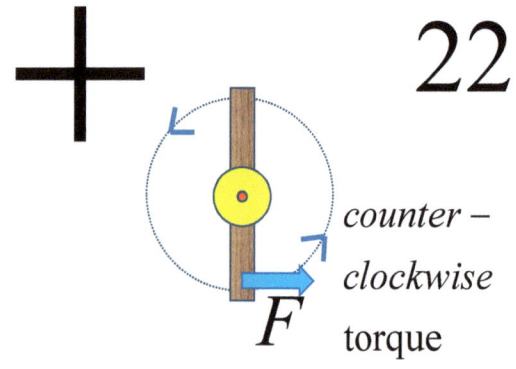

22

counter – clockwise torque

22

Two torques in opposite directions cancel to give zero net torque.

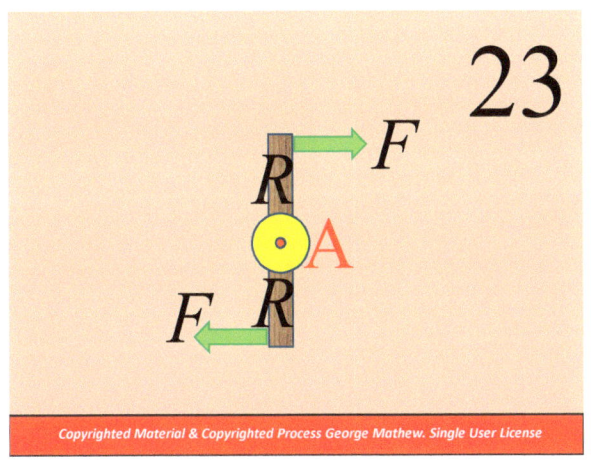

23

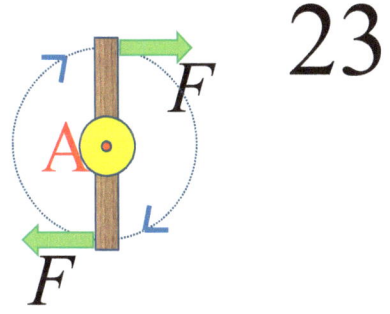

23

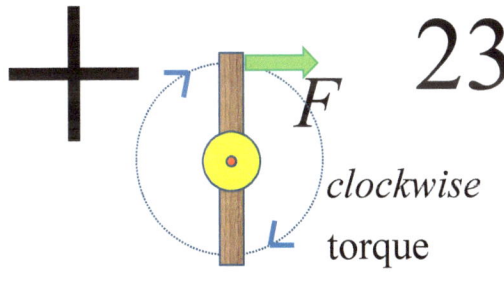

23

clockwise torque

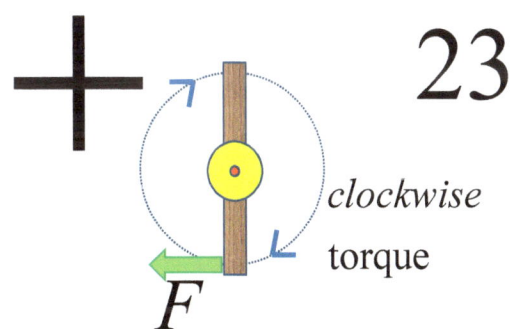

23

clockwise torque

Net torque is clockwise.

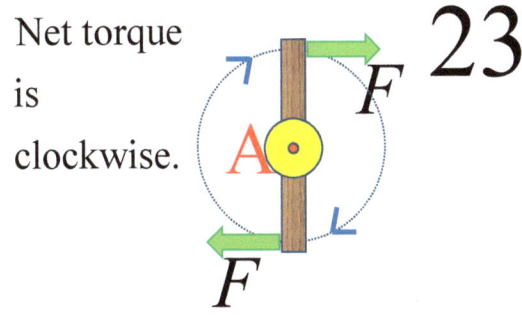

23

24

3/15/2013

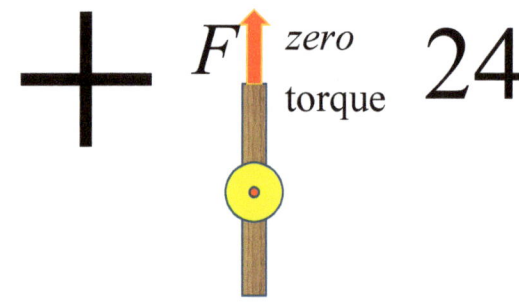

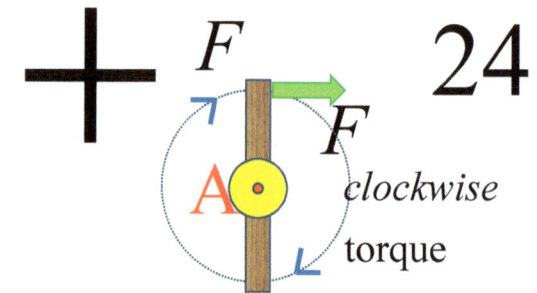

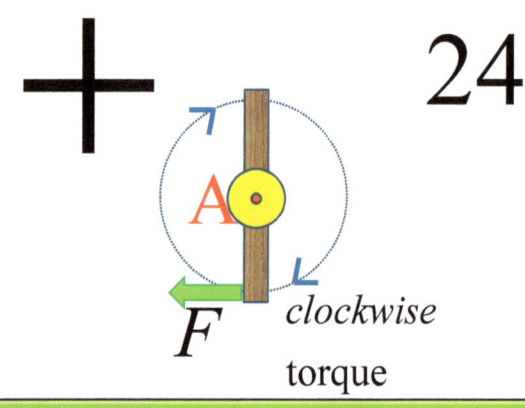

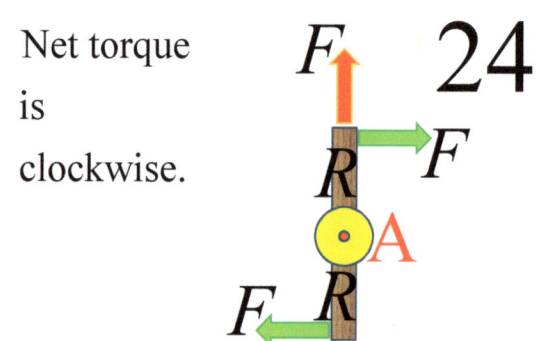

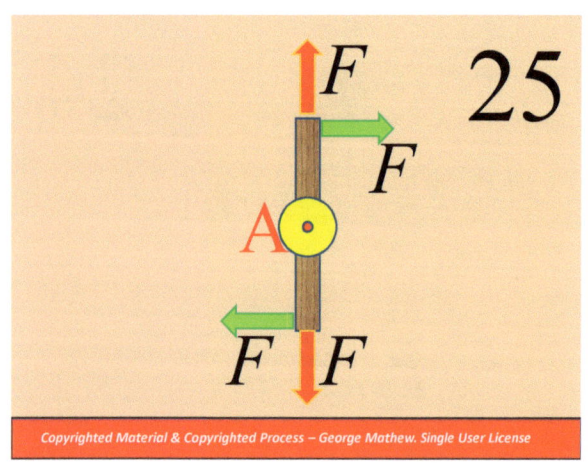

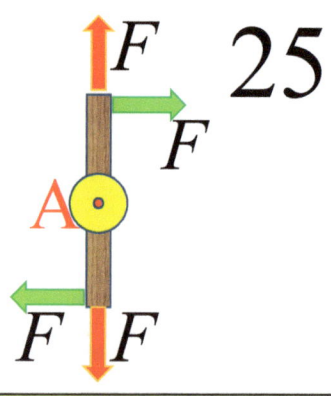

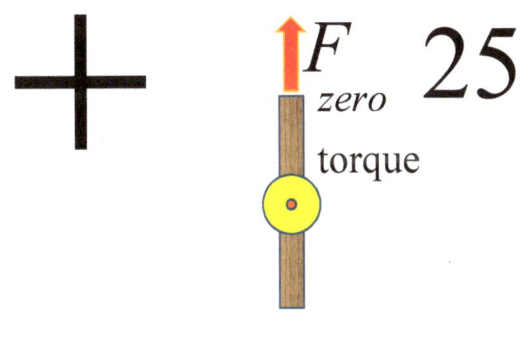

25

F
zero
torque

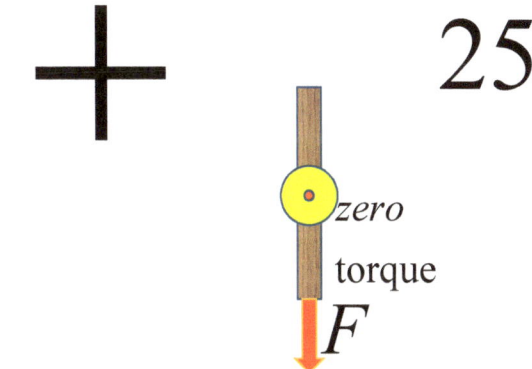

25

zero
torque
F

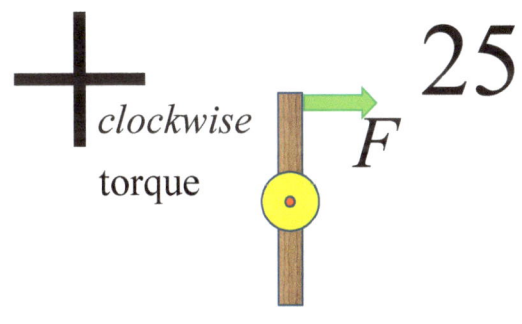

25

clockwise
torque
F

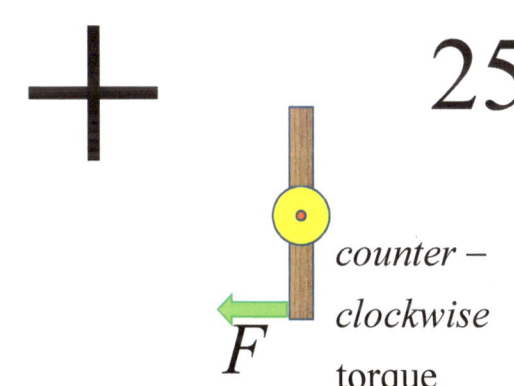

25

counter –
clockwise
torque
F

25

Net torque is clockwise.
(Both non-zero torques are in the same direction)

1. A
2. B
3. C
4. C
5. C

answers

6. C
7. A
8. C *answers*
9. B
10. B

11. A
12. A
13. B
14. C
15. A

answers

16. A
17. A
18. A
19. C
20. C

answers

21. A
22. C
23. A
24. A
25. A

answers

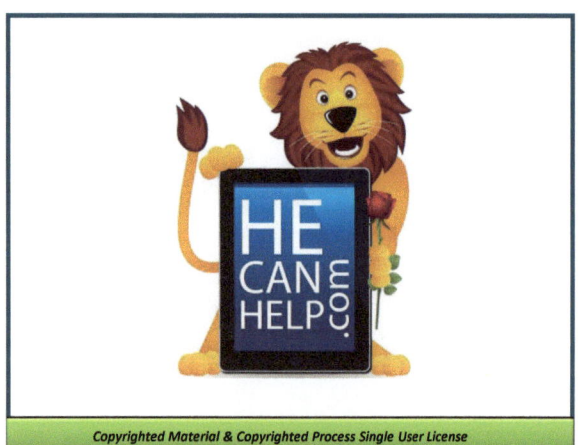

ACADEMICS ON LINE
PO BOX 1324
MASON, OH 45040
USA
info@hecanhelp.com

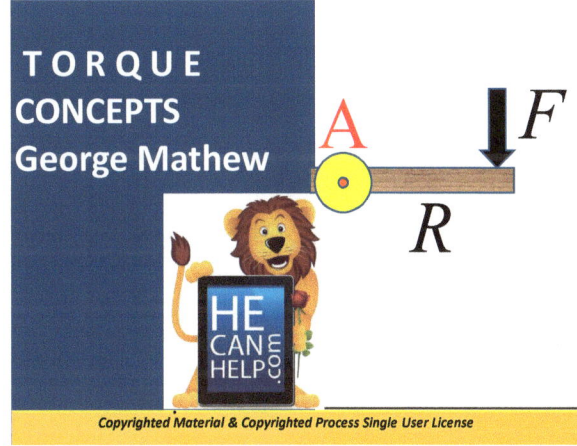

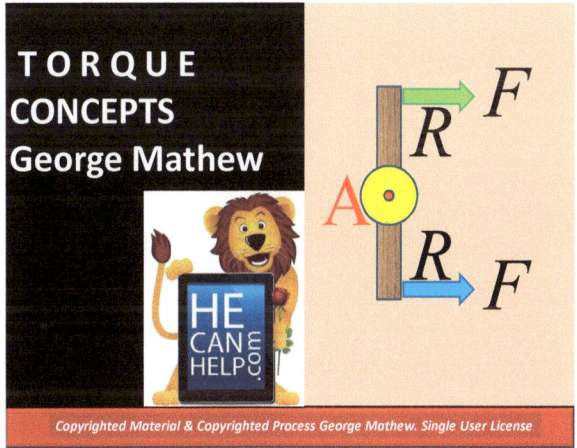